Quantum Gravity
A Study in Physics and Cosmology

by

Warner Schneider

PITTSBURGH, PENNSYLVANIA 15238

The contents of this work including, but not limited to, the accuracy of events, people, and places depicted; opinions expressed; permission to use previously published materials included; and any advice given or actions advocated are solely the responsibility of the author, who assumes all liability for said work and indemnifies the publisher against any claims stemming from publication of the work.

All Rights Reserved
Copyright © 2016 by Warner Schneider

No part of this book may be reproduced or transmitted, downloaded, distributed, reverse engineered, or stored in or introduced into any information storage and retrieval system, in any form or by any means, including photocopying and recording, whether electronic or mechanical, now known or hereinafter invented without permission in writing from the publisher.

RoseDog Books
585 Alpha Drive
Suite 103
Pittsburgh, PA 15238
Visit our website at *www.rosedogbookstore.com*

ISBN: 978-1-4809-7023-6
eISBN: 978-1-4809-7000-7

Preface

Albert Einstein determined that gravitation is the curvature of space in proximity to any celestial body of mass. If this is so, then how and why does it occur? And what is the relationship between matter and the fabric of space that allows for it? We can gain a more complete understanding of this force by identifying the various symmetries and field vectors that are replicated between the quantum level, the macrocosm of our universe, and the greater macrocosm lying beyond it. The dynamic underlying the expansion of space is revealed in this way, which goes a long way in explaining the origins of the universe and where it is ultimately headed.

QUANTUM GRAVITY
INTRODUCTION

We are fast approaching that time in postmodern physics when conventional science must acknowledge the new higher physics that seeks to incorporate, not only the physical, but all aspects of reality into it's overall picture. The door leading to that dimension may already have cracked a bit. Perhaps in the form of mathematical indicators or anomalous accelerator laboratory results that cannot be explained in conventional terms. One example of this is the action of virtual particles, which is erroneously explained as creation and dissolution. What is actually occurring is that they are moving in and out of our time continuum by means of a vector lying perpendicular to it. This action can be described as a rudimentary form of time travel, which is astounding in itself because it implies that we may be able to engineer the technology to travel time.

This work takes up and expands on the subjects that were first introduced in the book "Beyond the Universe" The final chapter 'Fact versus Belief' takes a humerous turn in adressing the subjective input that many members of the scientific community have in any field of study where one's 'personal bias' taints their research. Visual aids in cartoon form are included.

ONE
THE MATH

Even with all of the strides that have been made during the past century by means of the calculations and equations of the scientific method, there are still those who question whether mathematics are a reflection of the true nature of the cosmos, or a manmade contrivance that can only give a vicarious understanding of it's observable manifestations. Absolute homogeny would give the fabric of space infinite complexity or at least the potential for such and also precludes the notion that it will conform to any rigid numerical paradigm. Albert Einstein's descriptions of curved space (gravity) is the first indication that that fabric manifests any defineable and finite characteristics. Not surprisingly then, a certain point is reached where mathematical assimilations used in formulating postulates and theories of the universe begin to trail off in accuracy: no longer reflecting the phenomenon they are intended to describe. This is one of the reasons that a unified field theory has not yet emerged from those efforts, and also why general relativity cannot be brought into agreement with quantum mechanics. After all, the dynamics described by general relativity and quantum mechanics work in perfect unison with each other in the real world. It is only the theorists and their equations that end up at odds with each other.

There are a number of interactions and fluctuations seen in the universe that fall between the cracks of the strictest mathematical parameters. The relationship between the radius and circumference of any circle is a specific ratio. Yet it comes out as an infinitive when calculated mathematically. This is one example where the math fails it's "real world" description. The line paradox states that there are as many singularities on this half inch line — as there are in the entire universe. It must be remembered, however, that a singularity is nothing, nothing at all. Therefore, attempting to calculate the number of nothings over here as opposed to there is a futile exercise that will see the persuant's head making contact with the rubber room wall before the true answer is given in mathematical terms. Perhaps a better means of revealing this type of mathematical inconsistency would be in calculating infinite mass. Say that we have two spherical objects, each composed of infinitely dense matter. Any increase in volume of one object over the other will correspond to an infinite increase of mass for the larger object. But because there is no smallest increment of volumetric increase, the equation describing it could go on forever and still fail to reveal the mass differential involved, infinity times infinity times infinity… What this all boils down to is that at the most fundamental level, underlying it's matter-energy, the universe will manifest in abstract and intangible ways. Although the math is exceedingly precise in it's descriptions of the manner in which objects behave within a gravitational field, the specific geometry and way that the force manifests will defy the formal structure of mathematics written as equations. One of the reasons for this is the intangible effect of time incursion; as well as the apparent contradiction we run into when attempting to define the time dimension in terms of an undiminished state as opposed to the variable manner in which it transpires under the effects of velocity and gravitation (time dilation).

Quantum gravity, as well as the downward force we are all familiar with in our daily lives can be defined through a series of entanglements involving time incursion, the residual effect, and retrocausality, etc. The twelve defining characteristics are as follows.

UNDIMINISHED CONTINUITY

Suppose that a tourist aboard an airliner is winging his way from Chicago to Orlando. There are two facts that he can be absolutely certain of. First, he knows that Chicago hasn't mysteriously vanished upon departure and that, secondly, the city of Orlando is already there, in it's entirety, prior to his arrival. What this indicates is that because any location in space is not diminished by the perspective of the observer, space must then evidence the characteristic of undiminished continuity. Albert Einstein determined that space and time are so closely linked as to be inseparable. If his space/time connotation is to be taken literally, then what applies for the directional member of the partnership will hold equally true for it's durational counter-part. In other words, the fundamental characteristics of one translates equally well to the other. Thus, if we substitute 2010 for Chicago, the year 2020 for Orlando, and time traveller for tourist, the analogy becomes perfectly clear. Time is no more diminished by the perspective of the observer than space is. If time did not manifest undiminished continuity past, present, and future then in effect it would have to create itself and erase itself as it goes; a premise that becomes implausible upon closer examination because two fundamental aspects of gravitation will be eliminated along with it. Not to mention each and every time travel destination.

TIME DILATION

"For every action there is an equal and opposite reaction". This principle of physics is generally used in reference to directional-directional interactions such as billiard balls at the break. One step removed from it will be the directional-durational effect where, as velocity through space increases, time slows proportionately (time dilation). Yet this implies a directional-durational asymmetry in the universe that would be similar to stating that the teeter totter only tips one way. What is needed here is a durational-directional counterbalace; the equal and opposite reaction to equalize the time dilation effect. This is found in the principle of *undiminished continuity* which states that all moments of time past, present, and future exist in a state of *variable simultaneity*. Say that you pull a one hour motion picture film out of the projector. That hour long interval now becomes instantaneous, whereas it's cause and effect sequence (motion) has come to a halt. In other words, increasing the value of present time (prime moment volume) will act to step down velocity to the point of absolute rest. The analogy is quite accurate in its real world description.

Undiminished continuity indicates a dimensional thickness in it's description of present time. If the time continuum exists in an undiminished state then there will be no such thing as true present time because the instant we think of as the now would be relegated to the durational singularity that divides the continuum into past and future time. Therefore, the value of present time must be redefined in terms of a durational volume that is derived from equal measures of both the past and future. This value is referred to as *prime moment volume*.

Where cause and effect (motion) is the secondary characteristic of time it's primary characteristic will be that of undiminished continuity, which in various ways acts as the other side of the coin where time dilation is concerned. Although not obvious from our perspective, the flipside of time dilation is just as real and indispensable to the workings of the cosmos as the A side. This dynamic is one of the keys to understanding how gravitation manifests and why it does what it does.

Say that a thoroughbred jockey on his mount is in a race on a one mile track. Running in the middle of the field he sees a number of horses up ahead outpacing him while the remainder of the field lags far behind. The one thing for certain here is that regardless the rate of acceleration, de-acceleration, or overall speed of each entrant, they will all travel exactly one mile. What this indicates is that where the directional aspect of the cosmos is concerned, space (distance) has an absolute value but the speed at which an object travels through a given distance does not. Where time dilation is concerned we have to be more specific in describing the variables involved. Undiminished continuity indicates that time does have an absolute, immutable value. It is *duration* in time that does not. Therefore, duration is to time what velocity is to space. Once we examine the specific interactions between space-velocity and time-duration an overall dynamic begins to emerge. This can be described in terms of a compensation factor between the directional and durational aspects of the cosmos.

CURVED SPACE

If gravitation is the curvature of space in proximity to any celestial body of mass, then how and why does it occur, and what is the relationship between matter and the fabric of space that allows for it? Thus starts the series of questions that must be answered before we can arrive at a complete theory of gravitation. In figure 1 we see a straight object of matter. A similar object in figure 2 has been curved. It's ambient space will remain un-changed, however. There are two basic reasons for this. First, because of the minimal interaction between the molecular structure of the object and it's surrounding space, space will move through the object as the object moves through space, whether we bend the object or put it in motion. Secondly, it takes the entire mass of the planet to induce the curvature that is manifest around the planet. The effect of any small object on that space will be infinitesimal for that reason. If it were as easy to curve space as it is to distort the object in figure 2, then the energy crisis will have been solved, because you had simultaneously created a vector angle in the fabric of space which, for all intents and purposes, acts like a perpetual motion machine.

VECTOR ANGLE

As intangible as it might seem there are certain characteristics that can be deduced about and ascribed to space. Because it is subject to gravitational curvature it will have a greater substantive value than the mere length, width, and depth of it's measurements; meaning that although you may be able to draw all the air molecules out of a given area of space, the action will not create the absolute vacuum we believe, but a relative vacuum instead. This substantive value is referred to as the fabric of space, and it will have a calculable equivalence to the matter-energy that emerges from out of and exists within it; similar to $E=MC^2$.

Generally we are told that objects move through a gravitational field by following the curvature of space, but this is not entirely so. Satellites in earth orbit follow the curvature of space, whereas, an object in free fall through the atmosphere is following the *vector angle* of gravitation The vector angle always lies perpendicular to the curvature of the field. If the in-active medium of isotropic space takes on curvature, it then simultaneously assumes the dynamic of a vector angle. This is roughly similar to the way that a drawn bow appears, figure 3, and the manner in which energy is released. The difference between them being that they release it in exactly opposite directions. For gravity this is indicated by the arrow in figure 4. Also, the bow becomes inert once the string is released and it's energy spent, whereas, the gravitational vector angle is like a perpetual motion machine. We are under constant acceleration even while sitting at rest on terra firma.

The bipolarity of electromagnetism is easily identified because it is primarily directional in nature. What this means is that both the positive and negative poles of this force are found in space. An entirely different paradigm is indicated for gravitation, however. When we take only the directional aspects of this force into account, then it appears to be a monopole aligned to it's vector angle which manifests solely as the force of attraction, to the exclusion of any repulsion. In order to identify the positive pole of gravitation, therefore, we must look to the time dimension.

Every celestial body of gravitational mass will push on the *back door* to varying degrees, depending on the amount of mass that it contains. Evidence of this is indicated by the slowing of time on all such bodies. The back door may be defined in terms of the past, past time, a portal into the past or a worm hole. Of the hundred billion galaxies found in the universe, none are allowed past a certain point in the accretion of matter. Once a black hole reaches it's critical mass *time incursion* becomes extreme enough to initiate time reversal, which may also reverse the action of accretion into dissipation in maintaining directional-durational symmetry. This may describe what happens within some quasars, where an old galaxy has, in effect, recycled itself and been reborn as a new galaxy in the past. Indications are that worm holes associated with black holes can only lead into the past and that galaxies, which are all centered around supermassive black holes, have a certain maximum size that they cannot exceed without initiating that process.

We can define the closed loop interaction of electromagnetism by observing the pattern that iron filings make around a bar magnet. Though both are very similar the schematic is not so easily discerned for gravitation because it plays out over the entire time cycle of the universe before seeing completion. This encompasses billions of years, of course.

You'll recall Einstein's hypothesis for the excess radius of any body of gravitational mass. The effect is very similar to the whirlpool of water going down the drain once the plug has been pulled; except that the vortex configuration is manifest by planetary mass even while at rest, with the plug securely in place. What this reveals is the potential for time travel.

Because every body of gravitational mass pushes on the backdoor (past time) an object freefalling in the atmosphere could literally pass right on through to the other side, if it can avoid crashing into that solid object (the planet). Thus, in order to travel time, the vector angle of the field must either be diverged from the center of mass or artificially induced. Coincidentally, when riding a beam of light you'd run up against nature's other stopgap against time reversal, the speed of light constant. Paradoxically, that solid object always seems to be there stopping or slowing us down, whether it is or not. Thus, in order to avoid the natural barriers of the universe, the time travel impulse must be induced more or less perpendicular to the time continuum. This is what would break *time symmetry* and open the portal to another time or parallel universe.

The fabric of space may have an underlying complexity like layers of an onion. Not thin layers but, rather, where each one is another counterpart of the whole. In this model the force of gravity would be compounded as a succession of these counterparts assumed the verticle and horizontal geometry of the vector angle and curved space. Oddly enough there may be little or no curved space between two rotating black holes. Here, it is possible that the fabric of space has aligned itself exclusively to the verticle geometry of the gravitational vector angle.

TIME INCURSION

It can be stated that when the fabric of space moves from an isotropic to a geometric state of dispersion then a force of some sort will have been created, whether this occurs macrocosmically or at the quantum level. Where gravitation is concerned a key element of the manner in which it manifests is often overlooked. This is defined as the geometry of *time incursion*, and it can be described as follows. Say that you have a two dimensional plane, shown in figure 5. As long as it stays perfectly flat it will remain just that; two dimensional. The one in figure 6 is distorted along it's two dimensional axis with a resultant decrease of measured perimiter. Although it has no volume in and of itself, the plane in figure 7 curves into the third dimension of space and, thus, evidences a space incursion. If we substitute space for the two dimensional plane and the fourth dimension for the third, then the analogy becomes clear in that spatial incursion translates into a *time incursion*. The geometry of curved space indicates that a gravitational field is not limited to a simple three dimensional distortion, but also encroaches into the fourth dimension of time. This is evidenced by the slowing of clocks within any gravity field.

Time incursion is another one of the key elements in understanding the manner in which the force of gravity manifests. The process is described as being very exclusionary, however. Take any three dimesional object and no matter how you bend, twist, or distort it that action will not result in a time incursion. Although this phenomenon manifests as a result of the kinship between matter, energy, and the fabric of space, it doesn't work very well for objects on a small scale.

THE QUANTA FORCE

Given the complexity found at the quantum level and the uncertainty caused by occurrences at that level falling outside of any mathematical structure, a good question would be, why do we see such precise replication of subatomic matter and most specifically the energy quanta of our world. Without the cohesion of a specific force of nature being assigned to it how can the geometry of those assimilations be repeated in such precise patterns? The answer is found, in part, in the fact that the laws of physics are synonymous with the forces of nature. Neutrons, and quarks have the strong nuclear force to determine their structure and cohesion. Where light photons and energy quanta are concerned, that piece of the puzzle is still missing. what we are looking for is a fixed geometry that is immutable under replication.

The cornerstone of the $E=MC^2$ equation lies in the equivalency between matter and energy. You can't go from none to some or some to none in terms of mass differential or any force of nature when crossing from one side of that equation to the other; or else you're playing hocus pocus with the universe. In other words, the mass for energy quanta cannot be calculated as zero and the strong nuclear force must find a more fundamental counterpart of itself that go to make up all neutrons, protons, and quarks. This is defined as the *quanta force*, the fifth force of nature; the missing piece of the puzzle. Without it's binding action and cohesion light photons and energy quanta, which are described as being similar to a bundle of nerves, would simply dissipate into the fabric of space and cease to exist. Neither would light speed be constant without the defined vector angle of that force. The quanta force is also one of the key elements of this particular theory. If matter distorts the fabric of space then there must be an interactive link between the macrocosmic force of gravitation and the strong nuclear force, which acts on all matter at the quantum level. In other words, gravity must be unified with the strong nuclear force, and the quanta force allows this to be done.

RESIDUAL STATE OF THE QUANTA FORCE

Light speed is described in terms of a cooperation between each photon and the fabric of space. Evidence of this is seen in the curving of light around strong gravitational fields. If light existed and acted independantly of that curved space, then it's trajectory would not be affected by any gravity field. The principle of undiminished continuity determines that all moments of time, past, present and future exist in a state of variable simultaneity. Therefore, that beam of light will be at a virtual state of rest even while it is speeding along at 186,000 miles per second. Coincidentally, when standing on terra firma gravity determines that we are under acceleration even while at rest.

Kinetic energy (acceleration) can be traced directly to the quanta force dynamic. According to undiminished continuity all matter-energy will have a linear extention through the time continuum. Because curved space (gravity) is not merely a simple three dimensional distortion of that fabric, it will evidence an encroachment into the fourth dimension (time). It is this complex space/time geometry or, *time incursion*, that introduces the *residual state of the quanta force* into the picture by tapping it's durational extention through the time continuum, and specifically into the past.

We can't tie the strong nuclear force (matter) to gravity without the intervention, or intermediate action of the quanta force in it's residual state. What emerges from this series of directional-durational fluctuations is the perpetual motion of gravity. A reciprocal effect is also indicated between the vector angle and curved space which acts to compound the force of gravity in subtle, yet exponential ways. For example, an astronaut in earth orbit may feel no gravitational pressure on his body, yet the massive body of our moon, some quarter of a million miles farther out, is held securely in orbit. This cosmic oddity can be explained as a sort of cumulative effect due to a quantum entanglement of all objects and particles of matter within a specific gravitational field. Also, it must be remembered that space is a dynamic, not an inert medium. After all, if one's body is ripped apart in proximity to a black hole, it is not the collapsed star but, rather, it's ambient space that is doing the damage.

THE RETROCAUSAL EFFECT

Retrocausality sums up one aspect of the way that gravity manifests in the universe. Take the example of the donkey pulling on a leadrope. The farther back the beast pulls on the rope the more pressure there will be exerted on it. This analogy defines the interaction between time incursion and the quanta force. The more that space is curved, the greater will be the effect of time incursion in tapping into the residual state of that force, which manifests as the increase of gravitational pressure.

DIMENSIONAL THICKNESS

Although the page that this theory is printed on mimics a two dimensional plane, if it did not have the depth acquired from the third dimension, then it would not exist other than as a hypothetical singularity. In order to be part of that interactive environment described as the "real world" it must have *dimensional thickness*. Similarly, when we move from the second to the third dimension it will be found that space is not a time singularity; and that the differential between the directional and durational aspects of the cosmos does not come out as zero in mathematical terms but, rather, as a defineable ratio similar to $E=MC^2$.

THE CRUSH FACTOR

In figure 1 we see an object of solid matter with no distortion. A similar object in figure 2 has been curved. It will thus tend to crush at the bottom of that curvature and pull apart on top. This action is referred to as *the crush factor*, and it has some important implications in the way that gravitation manifests. In addition to the length, width, and depth of space, it's defineable characteristics also include curvature, vector angle, and time incursion which, in unison, act as the force of gravity.

Say that the two dimensional plane in figure 7 represents space and it's curvature is seen as a time incursion. This configuration would evidence no crush factor because it lacked the dimensional thickness taken from time that is required for it to manifest. Take away the crush factor and the real world result would be to erase the vector angle from the fabric of space and the gravity along with it.

The plane in figure 8 does have the necessary thickness and, therefore, will evidence the crush factor, which emerges as the vector angle of curved space. The indication here is that space is not a time singularity, and that the individual aspects of this dynamic always act in unison. Of course nothing is being crushed (due to the crush factor) because we're talking about the fabric of space. Only that that fabric assumes the specific geometry associated with the vector angle of gravitation. Evidence of this is seen whenever an object is put into motion within a gravitational field. If the fabric of space was in an isotropic state then the object would remain at rest, obviously.

It is possible to correlate the specific interactions between the defineable characteristics of space. In simple terms, we wouldn't have gravity without the vector angle and there can be no vector angle without the crush factor, which could not manifest without curved space and dimensional thickness. If it wasn't for undiminished continuity there would be no residual state or dimensional thickness. To sum up the chain of interactions that emerge as the force of gravity one might state that "all the dominoes either stand or fall together."

KINSHIP-REPLICATION

There are no two things in the universe that are absolutely alike or absolutely different. For example, a diamond is precious, beautiful, and valuable, whereas, that lump of coal is ugly and virtually worthless. Yet both objects are composed of the same kinds of neutrons, protons, and electrons. On the other hand it would be safe to say that there are no two neutrons to be found in the universe that are exactly alike, if for no other reason than that no two can occupy the same space at the same time. Even the same neutron is not exactly alike from nanosecond to nanosecond. If we consider the equation $E=MC^2$, then it becomes apparent that each one is comprised of an exponential amount of energy, meaning that it will be virtually impossible for a neutron to be the mirror image of any other neutron for more than a nanosecond. In other words they are all the same yet different. We are all familiar with Brownian motion and the crystaline structure of quartz, where similar patterns and motions are replicated and repeated over and over again on both the small and large scale. This is very much what we see at the planetary scale where the earth is curved, therefore, it's ambient space is also curved (gravity). The overall paradigm that emerges is thus defined in terms of *kinship and replication.* If we ran a DNA test on matter, energy, space, and time then the results would come back positive, indicating that they are like first cousins to each other. In fact we may be able to employ the initial equation in order to describe the specific mathematical ratios between the four most fundamental aspects of the universe. Here, M is matter, E is energy, S is space, and T stands for time. If $E=MC^2$, then $S=EC^2$, and $T=SC^2$.

The specific ratios notwithstanding, we see a certain comonality between all participants inspite of outer appearances. You may recall the comedians The Three Stooges. They all behaved similarly because of their similarities. After all, it was Three Stooges not two Stooges and a giraffe. This goes part and parcel with the hypothesis that the universe is not a random collection of junk that behaves in a haphazard manner but, rather, that it acts in a cohesive, purposeful way as a single entity, because there is no "odd man out" in terms of its matter-energy, space/time, or any force of nature.

STANDARD DURATIONAL COVARIANCE

A two dimensional plane requires a certain amount of "thickness" derived from space to elevate it from that hypothetical status and allow it to interact with the real world. A similar relationship is revealed between the third and fourth dimensions of the universe. Where dimensional thickness infers that space is not a time singularity, *standard durational covariance* will reveal the specific ratio between the directional and durational aspects of the cosmos in mathematical terms.

There is a compensation factor at work between the space and the time elements of our world that will not allow it to be defined in absolute terms such as nothing or infinite. Although cohesive, the quanta or photon is not a singular, simple unit of absolute uniformity channeled into a nonconflicted vector. Instead, the push, pull, and tug inherent to a unit described as being similar to a bundle of nerves, will see it's clock nudged off the position of absolute rest; and determines that light transpires at the least allowable increment of time. This interval may be corrolated to the dimensional thickness that space derives from time and will, thus, prove useful in determining the standard durational covariance of the universe. Where light speed is concerned, the durational interval referred to as the second is based on an arbitrary system of measures, as is the 186,000 miles that light travels during that interval. These must first be reduced to their lowest common denominators in terms of their real world corrolation, before we can make a specific calculation of so many spacial miles being equivalent to X amount of time. There are three indicators that may be used in that regard; the durational interval of virtual particles, light speed, and the equation $T=SC^2$. These are just indicators, of course, and lack the degree of certainty to act as proof.

In figure 9 the plane represents space with a beam of light (the arrow) passing through it. Here, if space was a time singularity then light speed would appear to be instantaneous, because it transpires at the least allowable increment, which corrolates to the dimensional thickness of space. Virtual particles would be undetectable for the same reason. If we add the dimensional thickness of time to the directional aspect of space, as seen in figure 10, then light speed will be stepped down to a finite, measurable velocity, 186,000 miles per second to be exact. *Variable simultaneity* indicates that although time itself exists in an undiminished state, it does not do so in the manner of a freeze frame; where people and events are eternally frozen there for the traveller to view at his liesure. Should you travel back in time to Dallas Nov. 22, 1963 in order to study the Kennedy assassination, you will not find that second shooter on the grassy noll, the Carcano rifle, or Abraham Zapruder filming the presidential motorcade. Dallas itself might still be there in some form or other but president Kennedy is not. He is gone. Time travel logistics are not what we might imagine them to be, and just as the contents of space are subject to endless variation, so are the events in time. The very act of travelling time would be an interruption in the cause and effect sequence of the action itself, like a contradiction that had just contradicted itself. Should the traveller arrive safely at his destination, only then will he be allowed to transpire normally again, according to the laws of physics. The principle of *sequential multiplication* states that you cannot go back in time to relive an era of your past over again (the first time). You can't relive it the first time twice, the first time three times. You can only live it the first time once, the second time once, the third time once... This may be something that is just too obvious to figure out, but it makes the grandfather paradox a proposition that is out of durational sequence for achieving the desired results; similar to shutting the barn door after the horse had escaped.

The T in the equation $T=SC^2$ refers to the *durational* aspect of time, which is a *variable* commodity. The standard durational covariance of the universe will most likely be based on the constant value of time, as defined by the principle of undiminished continuity, otherwise it would fall squarely into the sliding scale category. Quantum gravity is one of the prerequisites of interstellar and time travel. Should man develop this technology, then the equation describing standard durational covariance will be as indispensable to the time traveller as $E=MC^2$ is to the nuclear physicist. It determines that you may be able to travel vast expanses of a lower dimensional state (space) by crossing relatively in-signifigant measures of a higher one (time). In other words, light years of space are equivalent to seconds of time. The dynamic does not violate but, rather, multiplies the speed of light constant.

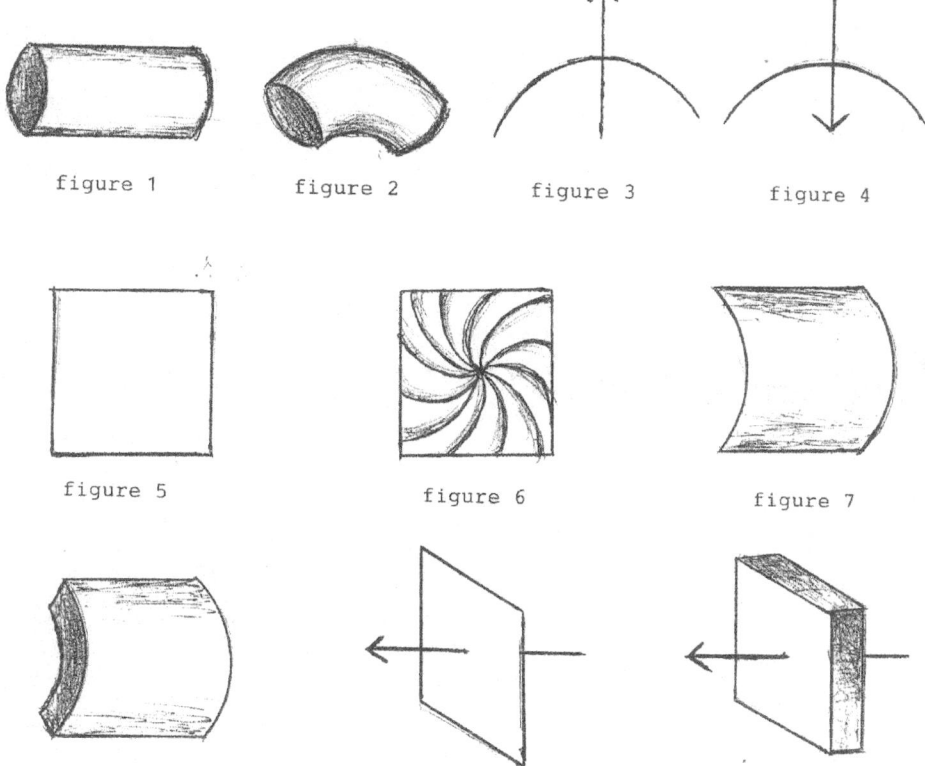

TWO
DARK MATTER, DARK ENERGY

We are told that 90% of the mass in the universe is missing. Before searching for or identifying this "dark matter" perhaps we should define or redefine the meaning of "mass" at the most fundamental level. Gravitation is defined as the curvature of space in proximity to any celestial body of mass. This indicates that there is a commonality and kinship between matter and the fabric of space itself which acts to link all aspects of the cosmos together including it's matter, energy, space, and time. If this is so then the characteristic defined as mass will translate and be inherent to all aspects of the cosmos, Higg's bosons and gravitons notwithstanding, The universe, in other words, is not a collection of miscellaneous junk behaving in a haphazard manner, rather, it is a single entity and acts as such. The interaction between matter and the fabric of space (gravity) determines that space has more substantive value than the mere length, width, and depth of it's measured volume. The requirement of a "Higg's boson" is, therefore, as pointless as looking for an extemporaneous agent in explaining the characteristic 'wetness' of water. Water is wet in and of itself period. No 'wet boson' is needed.

Einstein said that God doesn't play dice with the universe. Neither should science play hocus pocus with it. If any mathematical computation gives a nil or infinite result then it is in error. For example, you can't go from none to some or some to none in terms of mass differential or force of nature when going from one side of $E=MC^2$ to the other, without contradicting that equation. All matter can be converted to energy. If energy quata are further disseminated into the fabric of space their mass doesn't go to nil but, instead, corresponds to a proportional expansion of that space, likely in accordance to Einstein's ratio. Thus, if S stands for space, and $E=MC^2$, then $S=EC^2$. When following this line of reasoning the best estimate is that the mass of the space in the universe is equal to the mass of the matter. When taking into consideration the 'free' energy floating around the cosmos then we've more than doubled the mass that has been accounted for.

It is possible that black hole matter has a way out of it's dead end state other than through the infinitesimal effect of Hawking radiation. Here, once a certain critical mass is reached within a supermassive black hole time reversal kicks in and opens a cosmic wormhole to the black hole's counterpart, a white hole where matter-energy is ejected. Wormholes are not aimed at arbitrary locations in space. Instead, gravitation's positive pole leads directly into the past. The farther back in time such a black hole is stretched the greater this reach will extend in it's present day galaxy. You might think of it as the "bungee cord principle" plus a bit of quantum entanglement thrown in for good measure. The requirement of durational-directional symmetry determines that any white hole action must have a gravitational rebound effect in it's precursor galaxy which is in excess of that predicted according to the mass of that galaxy and will, thus, account for much of the gravitational effect that has been attributed to dark matter. Also, there may be a gravitational effect between parallel universes depending on their durational proximity from each other and their arc of spatial expansion as shown in figure 3.

In adressing the question of "dark energy" it must be remembered that the fabric of space is not an inert vacuum but a dynamic interactive medium having commonality with the matter-energy it contains in our universe. There is no missing agent or additional element yet to be identified that permeates it, only the medium in and of itself. When taking into consideration the transigent action taken by the fabric of space in facillitating the expanding universe then the dark energy mystery is most likely solved.

An object will be torn to shreds in proximity to a neutron star. It is not the star itself but it's ambient space that does all the damage. This dark energy inherent to the fabric of space may be the most powerful force to be found in the universe.

THREE
THE BIG BANG

To bang or not to bang that is the question. The most fundamental question in understanding the pathology of our universe. At the time of this writing the big bang theory is almost universally accepted as proven fact in scientific circles. And when looking at the evidence on the surface, as presented by conventional science, the event scenerio of that bang seems to be irrevocably certain. Even the book "Beyond the Universe" had been written in it's entirety under the assumption that the big bang theory was correct. A signifigant error, as it turns out, in laying the foundation for "Counterpart Theory". As indicated by the distortions of gravitational geometry, the substantive value of space is far from being an 'absolute vacuum'. Light speed can be defined as the cooperation between light and the fabric of space. The evidence for this? Like a train on curved tracks, light follows the curved space geometry of a strong gravitational field. Also, space is a dynamic, interactive medium that can tear an object apart in proximity to a large celestial body. It proves to be an incredibly powerful force upon having it's vector angle increased.

It has been estimated that the universe has enlarged fifteen percent during the last billion years. If that is due to the expansion of space, then the question must be put forth, how much can you thin out the fabric of space before that beam of light would have nothing to latch onto, and move along? A slight expansion of space might actually facillitate an increase in the speed of light constant but put in an absolute vacuum devoid of that fabric, light speed would be reduced to the state of absolute rest. If the fabric of space is expanding then the substantive value of that fabric is being progressively reduced towards the end result of becoming an absolute vacuum.

If the speed of light is to remain constant then the fabric of space must remain uniform and constant also. Upset the balance between light speed and spatial uniformity and we might as well throw the laws of physics out the window too and watch as our structured orderly universe enters a state of absolute chaos. This is a major problem for the big bang theory that remains unanswered.

If we were to reverse time for the estimated age of the universe (13.5 billion years) it is believed that the cosmos would be reduced to a singularity of infinite mass. This is exactly the way things must play out unless an underlying dynamic is identified that acted to interrupt the sequence of events related to both expansion and contraction. By studying the action of standard candles in other galaxies it has been determined that the expasion rate of the universe is actually increasing at this time. Should this runaway action continue unabated, the cosmos will disperse itself infinitely and cease to exist. In order to eliminate the infinities of endless dispersion and infinite mass, the universe must be subject to a 'normalization' that precludes it from destroying itself. The ancient Vedic texts refer to this cosmic action as "God breathing". Coincidentally, when one breaths they do not do so to the point of either bursting or collapsing their lungs. If our universe is to be a fixture within the greater scheme of things for a longer term than cosmologists predict, then it will have to evidence an action very similar to the Vedic description of God breathing.

Albert Einstein determined that space and time are so closely related as to be inseparable. Therefore (as previously stated) the fundamental characteristics of one will translate equally well to the other; the first of which is undiminished continuity. All locations in space exist in a simultaneous state, therefore, all moments in time must do the same.

Undiminished continuity determines that the time dimension is a vast expanse encompassing all eternity, parallel time lines, and higher dimensional worlds.

Because any volume of space is defined by the three dimensions of length, width, and depth; time must also have three corresponding dimensions. These are continuity, transigence, and time radius the fourth, fifth, and sixth dimensions of the universe. Just as the first dimension defines spatial length, the fourth dimension (continuity) defines durational length. The second dimension gives us directional width, while the fifth dimension, transigence, gives us durational width allowing our time line to expand transversely into the greater physical time plane. This is where countless other universes will be found that lie durationally parallel to our own. Where the third dimension defines spatial depth the sixth dimension (time radius) is what gives us durational depth and allows for a multiplicity of time planes beyond the physical one. This is where the higher dimensional worlds are found. In the physical time plane both the separation between, and common denominator linking parallel universes to each other will be the shared space intrinsic to all of these worlds. In this model the expansion and contraction of space is not a self contained action within any single universe. Rather it is a universal dynamic that is shared between all worlds within the time plane. Here, the fabric of space will manifest a rising and falling action similar to waves on the ocean. There is no break or discontinuity between individual waves on the sea, thus allowing the greater body of water to act as a single entity. So it is within the greater macrocosm and it's shared medium (the fabric of space). The third and fourth dimensions describe the immediate macrocosm; our space/time. The fifth dimension thus defines the "greater macrocosm" of the physical time plane.

Because the expansion rate of the universe is increasing it can be stated that we are on the crest of a wave at this time. Once the wave passes a contraction phase will ensue. A reciprocal action is indicated in the single overall confluence between neighboring universes where as some are expanding others must contract. Waves in the ocean do not rise so high that they go into earth orbit. Neither do they bottom out on the ocean floor. Similarly, if the fabric of space evidences this type of compensating action then universes within the time plane will not be predisposed to blowing themselves apart or collapsing in on themselves, and our universe will be far older than the estimated age of 13.5 billion years. This effect is referred to as 'normalization' and good evidence for it is that there are stars in the cosmos older than 13.5 billion years.

If a round cannon ball passed through a two dimensional plane it would appear suddenly as an expanding circle, then contract and disappear. What's missing from that perspective is the greater action occurring in the third dimension, the cannonball travelling through space. We encounter a very similar problem when attempting to understand what is happening to the fabric of space in terms of the expanding universe. This is the transigent action of space being exchanged between parallel universes in the fifth dimension; the greater macrocosm of the physical time plane. Physicists propose that virtual particles are created from out of space then dissipate back into it, thus, ceasing to exist when, in fact, there is no creation or cessation taking place here. No more than it did for that cannonball passing through the plane. That idea is simply a trick of perspective due to not being cognizant of what is taking place in the greater macrocosm. In that regard foreign matter travelling in the fifth dimension can pass through this universe by intersecting the narrow window of our time continuum in a perpendicular manner. The brief instant that virtual particles trespass in our world may be used in determining the ratio between the directional and durational aspects of the cosmos or it's "standard durational covariance". It's synonymous with 'dimensional thickness'. Virtual matter is swept along by the fabric of space in a cumulative action described as the expanding universe dynamic.

The butterfly pattern defining the curvature and vector angle of a gravitational field (figure 4) in the macrocosm is replicated in the greater macrocosm. Here, the curvature of spatial expansion is the horizontal, whereas matter-energy takes the verticle path described by continuity (figure 3). Also, we see reverse symmetry in that the time arrow is reversed 180° from the gravitational vector angle.

To sum up this study and reiterate the main points, the big bang is the most logical explanation of the universe when looking at it through the narrow window of our own macrocosm but from the perspective of the greater macrocosm it becomes implausible.

If a swimming pool is being filled with water you don't say that the water is expanding but, rather, that more is being added to it. Similarly, the dispersion factor and consistency of the fabric of space has always remained constant. It has never been subject to the extreme fluctuation described in the standard cosmological model. The universe ages vertically in accordance to the time arrow and expands horizontally as the fabric of space ebbs and flows in a wave effect perpendicular to the time continuum. This is the transigent action that defines the underlying dynamic of the expanding universe.

FOUR
STRINGS AND MEMBRANES

Proponents of the membrane universe theory propose that the three dimensions of space exist in the form of an immense membrane or 'brane. If you go beyond space as we know it there will be countless other such 'branes of enormous magnitude. The hypothesis fails to adress a number of contradictions, however. First, it is believed that when two membranes touch it will initiate a bing bang thus culminating in the creation of a new universe. Yet that idea violates the conservation of matter and energy. Einstein said that God does not play dice with the universe. For obvious reasons he will be unable to have played hocus pocus and created it out of nothing (membranes touching). Nothing touching nothing equals nothing. Although he might be able to play lotto with it providing that countless other members ('branes) have played and lost resulting in a cumulative big crunch. Thus contributing to the jackpot that he ultimately cashes in on; a new universe.

If these membranes really do exist out there somewhere, like flapjacks floating around untethered at breakfast time, proponents fail to answer one important question. What is it that separates them from each other? If a slice of lunchmeat were placed between each one we could refer to that configuration as the 'baloney sandwich model'. The bottom line here is that space can't go beyond space. You can't draw arbitrary boundaries in space (membranes) and say this is space and that's not space, this is and that's not etc. If you want to go beyond space you must look to the time dimension, but time takes space along with it.

A beam of light will follow the curvature created by a strong gravitational distortion. There will be no such effect due to the expanding universe action. Here, space doesn't move to the left or right, up or down because the expansion is not a directional but a durational effect described as a transigent action. It therefore will remain directionally isotropic and evidence no distortion or motion, which precludes any movement of membranes due to the expanding universe dynamic.

Next, we'll take a brief look at string theory. This is the offshoot of physics that has been touted as "the only game in town" when adressing the shortcomings of particle physics. The study has taken on monstrous proportions with tens of thousands of pages in calculations and numerous written materials coming out. Yet the idea is based on a false premise; that matter is composed entirely of tiny vibrating strings that have length only; linear singularities in other words. Even if these strings did exist as theorized they would have no effect on the environmet whatsoever. Singularities are not vectored or subject to any sort of assimilation in and of themselves. Instead, they are delineated from out of a three dimensional volume in a hypothetical sense only (in the mind of the one doing the theorizing) and, therefore, are not part of an interactive environment (the universe). As such, they exist exclusively in a virtual state. If singularities don't exist in actuality and matter is composed entirely of singularities, then matter doesn't exist either. By following this line of reasoning matter-energy, our world, and we ourselves can be whittled away to nonexistence.

You might say that string theory has a built in safety net of sorts. Bigfoot is easily enough debunked because if it can't be found then it, likely doesn't exist. Strings, on the other hand, were given an infinitesimal size of 10^{-35} of a centimeter making them too small to detect, and seemingly designed to be debunk proof; a strategy that is not an option for believers in sasquatch.

In dairy farming when one cow has been milked dry the farmer moves on to the rest of the herd. Where physics is concerned, the cow in question takes the form of extra spatial dimensions, a hypothesis that over the years has certainly been milked for all it's worth, bone dry in fact. It's very near the point where science needs to move on to the rest of their herd. These are the extra time dimensions and subdimensions.

String theory is correct in the number of 'extra' dimensions that remain unseen from our perspective in this world but as stated they are time dimensions not spatial ones. Neither are they coiled up infinitesimally as physicists had believed. Instead they are fully expanded and manifest as the higher dimensional worlds that extend above and beyond physical reality. It is only in this universe that they have a minimal connection. Think of a flat plane (representing the physical) with lines (the extra dimensions) intersecting it in a perpedicular manner. A being living on that plane would be unaware of the limitless extent of each line, only the infinitesimal connection they had to the plane. This is analogous to the erroneous conclusion that string theorists have reached concerning the manner in which the extra dimensions are manifest, and it is due entirely to the limited perspective that our world places upon us. In a certain sense we are as restricted as beings living on a two dimensional plane. It is an odd coincidence that string theory's mathematical calculations have indicated the same number of extra dimensions as had been revealed in the ancient Vedic texts pertaining to the higher realms of eastern religeon. If religeous belief allows a spiritual connection to one or more of these worlds then atheistic scientists and decidedly nonmaterialistic eastern mystics may compliment each other more than either would care to admit.

FIVE
TIME TRAVEL

If one were planning a trip to Paris there would be no doubt that the city is there intact in it's entirety ready and waiting for their arrival. This is because all locations in space exist in a simultaneous state. The possibility of time travel reveals the flipside of that correlation. You can't go someplace that doesn't exist, neither would a virtual or potential manifestation of the past or future be enough for for the traveller. Any destination in time must be as real and solid as Paris.

From our perspective here in the present, appearances indicate that time creates itself and erases itself as it goes with the past gone and the future existing only in a potential and not an actual state. If that were true then obviously there could be no such thing as time travel. The characteristic of 'undiminished contiuity' is, therefore, the pre-requisite for any such mission. Yet 'variable simultaneity' determines that moments of time do not exist in a freeze frame state, which would preclude backtracking on the continuum in order to study history or travelling foreword in time to predict the future.

Beyond that basic understanding of the time dimension we must also expand our time line into the greater time plane of the fifth dimension. A multidimensional time continuum is not a quantum leap in theory just the next logical step that must be taken by conventional science. This is how it works. Say that a man is standing on a set of tracks in the middle of a railyard containing countless other tracks. The last thing you'd expect him to say is "the railroad track that I'm standing on is the only one there is". He'd know better than this because, obviously, he can see the numerous other tracks. Not so where parallel time lines are concerned. We are able to see only one, the one that we transpire upon along with our universe. The reason for this is that the laws of physics governing our world do not allow matter-energy, or visible light to jump the tracks of the their home time line in order to enter a neighboring one. Thus, we cannot and do not see what is going on in other time lines or that they even exist. Because conventional matter-energy is only allowed to move in accordance to the time arrow of any univeverse and not in a manner perpendicular to it, the zone between parallel worlds must remain devoid of all discernable form, save for the fabric of space itself. If a time traveller ended up in that zone, other than what his own lighting apparatus emits, he will find himself in total darkness. He and his machine would, in effect, be a tiny self contained universe subject to it's 'own' time arrow and transpire independant from and parallel to the time arrow of his 'home universe'. No visual contact or radio communication could be made between the two as all such attempts can never be exchanged. Instead they must always run parallel to each other according to the laws of physics.

The potential dangers of "time travel entanglement" determine that you cannot backtrack precisely on the time continuum or else you may regress in age and perhaps cease to exist altogether upon passing your date of birth. The same applies for going into the future. Say that a man planned travelling fourty years into the future to witness his own funeral. He thus follows the continuum too precisely and instead of being the detatched observer there, mission control finds him lying dead in the coffin. For this reason time travel entanglement dictates that any time travel mission will be initiated as a transigent action that takes a roundabout course to it's destination.

If parallel universes have a durational kinship and entanglement then they will also be linked directionally according to the expanding universe dynamic. Einstein's space/time connotation indicates that you can't toss a coin without both sides of it having been equally tossed. Therefore, if the empty zone between two parallel universes measures a durational extent of one second, then there will be a corresponding directional width of 186,000 miles between the two. This in accordance to the calculated ratio of 'standard durational covariance'. The second is an arbitrary unit that man devised to measure time. Same thing for the mile in measuring distance. Once we have discovered the 'archetypal second' intrinsic to the universe, our second can then be adjusted and conformed to it, as well as it's directional counterpart for the mile. This will reveal the lowest common denominator for standard durational covariance and likely the duratioal proximity of our nearest neighboring parallel universe. If a perfect circle represents present time, then all material objects are perfectly centered within it. Certain subatomic particles may become loose within that designation and are thus subject to "time wobble" whereby they behave erratically within the present or the "prime moment" . This durational action manifests as an asymmetry. Under certain conditions a particle may break time symmetry and leave that durational circle or, the prime moment. This is the initial impulse of time travel where the particle is said to have disappeared. It is described in terms of a transigent action where the particle splinters off of the time continuum in a manner more or less perpendicular to it. Large material objects are not subject to this sort of action under ordinary conditions, of course, which is why time travel has not yet been undertaken by mankind.

Whether were talking about an airplane, rocketship, or a person taking a stroll through the park, all of the above are kinetically and conventtionally vectored, meaning that their motion is limitted to the three dimensions of space. You can point in any direction desired north, south east, or west, up, or down. But if asked to point to the past or future you'd be at a loss for another dimension and totally stumped. Neither could any kinetically powered vehicle take one there. This is where the added capacity of field vectoring comes in allowing one to break time symmetry to travel the fourth and fifth dimensions. Perhaps, surprisingly, it turns out that by pointing straight down you are indicating a durational location because as described in chapter one; every celestial body of mass pushes on the back door, with that being past time. You just can't get there because obviously you'd be crashing into a solid object; planet earth. Therein lies the clincher, you must be able to divert the polarity of gravitation in order to utilize it in field vectoring. Although we are able to detect only the negative pole of gravity in our world; it's power of attraction, it does manifest a positive pole. Gravitation and electromagnetism are both closed loop interactions as indicated by the compensation between opposite poles. The north and south poles of a magnet are both found in space. Only the negative pole of gravitation can be identified in space. However, it's positive pole manifests in time; the past. Because this force spans beyond present time it may be utilized to break time symmetry in order to facillitate time travel. But we are unable to do so because it cannot be controlled or directed. Electromagnetism, on the other hand, falls under our control but does not span beyond the present and thus cannot be used to induce time travel; a textbook catch 22 situation. The conundrum is solved by creating a secodary force through a coupling of the two primary fields. In this way we have generated a force that both spans beyond present time and is under our control and direction. A force field can be described as a geometrical distortion of the fabric of space. Macrocosmically electromagnetism manifests as lines of force between opposite poles. Gravitation is a bit more complex in it's emergent curved space configuration, which incorporates the vector angle and also time incursion. Obviously, the field geometry of both forces will become exceedingly complex at the quantum level, and this is exactly where the coupling between them must be initiated. Yet we expect these to conform to a specific paradigm and thus be defineable for both. Initially the precise mathematical model for each force must be described macro-cosmically and at the quantum level before the common denominator is reavealed that will allow both forces to merge and emerge as the secondary field we are seeking. It may be possible to construct schematics that can be overlaid to reveal the quantum field geometry for each field and at which juncture the coupling dynamic can be induced. We must then determine how to sustain and control that force.

Electromagnetism is a three dimensional distortion of space, whereas, gravitational distortion extends along no less than four dimensions. The three of space and one of time, in accordance to the description of time incursion. The dimensional disparity involved could inhibit the coupling between the primary forces making it exceedingly difficult. The countless specifics of that process and subsequent engineering of a functional technology described as the "field propulsion system" will be extremely complex, difficult, and beyond the scope of this study.

Conventional science still holds to the belief that it is impossible to travel faster than the speed of light and, of course, this is true for any kinetically vectored craft travelling through space. Standard durational covariance indicates a specific ratio between the directional and durational aspects of our universe where one second of time is roughly equivalent to 186,000 miles of space. So if field vectoring is employed to induce the time travel impulse, you are not violating the speed of light constant but multiplying it, which for all intents and purposes makes journeys of relatively close proximity instantaneous. In a manner of speaking to get there you already have to be there because the tidal forces and logistics of the universe won't allow you to do so incrementally. In other words, the inertial properties inherent to distance and duration cannot be overcome or violated only superceded. Einstein determined that time slows for an object as it's velocity increases. Also, there is a flipside to relativity where time acts to diminish velocity. Here, a certain critical point is reached that allows a flip flop of directional and durational vectoring where, if the prime moment is breached, time stops motion which then (paradoxically) becomes instantaneous, depending on the perspective of the observer. This is the principle that would be employed by highly advanced civilizations on their planet hopping missions throughout the galaxy.

'Field osmosis', or the saturation of an object with an amplified force field will act to break time symmetry, thus allowing it to travel the time dimension. Our time continuum is a linear extention with a durational width that falls within the descriptions of 'standard durational covariance' and following it precisely into the past or future would be as difficult as walking a balance beam. And the time traveller is most likely to find himself in the void between parallel universes, for that reason. Directional control will rely heavily on the 'path tracer' principle, which states that an object is most likely to retrace a previous path it had taken through space/time as opposed to a totally random course. This makes the return home (upon completion of a mission) easier and determines that all time travel vehicles must undergo a process of being broken in, and indicates that a machine slated for missions to Zeta Reticuli will undertake missions to Zeta Reticuli only and not to Mars or the battle of Gettysburg. The path tracer dynamic determines that after the completion of a mission each subsequent one will become more reflexive and automatic. There is a down side to that, however. After so many missions the cumulative effects of field osmosis prediposes a material object to become "loose" in it's position within present time, and subject to transigent action.

Having undertaken one too many time travel journeys, a craft will be put away in it's hangar. Returning next day to service it the crew chief is surprised to see that it had disappeared from it's moorings, vanished into thin air. Under normal conditions people, objects, and vehicles are perfectly secure within present time. It's much like sitting at the bottom of a ditch. You're stuck there and unable to move. But upon having come under the effects of 'field osmosis' present time inclusion can become quite precarious, even after power has been shut off; like balancing on a high wire where you're likely to fall off. This is why each time machine will be retired after a predetermined number of missions. For that same reason the crew of said ship must retire along with it, lest they inadvertantly find themselves lost in time. Also, upon returning home after every mission field polarity must be reversed for a specified time in order for the ship and crew to once again become centered within the present, and minimize the chance of remaining "loose" within that coordinate. Having reached the end of it's useful life, an interstellar craft or time machine must be put out of commision as a unit. You don't disassemble and use it in constructing a new one, as the disparity in resonance between new and old components will cause the new ship to self destruct during field amplification.

Every craft will reach a certain point where it may 'jump time' unexpectedly and will thus be labelled as hazardous material. Because field osmosis alters the electromagnetic structure of material objects the craft may pass through thin solid barriers such as doors and walls. But there is a limit to that, which is why the most expedient means for disposing of an obsolete craft involves placing it in an abandoned mine shaft. Should it jump time there, the deeper it penetrates into the mine wall the progressive increase of hinderance between the craft and the solid wall determines that the molecular structure of both will become enmeshed with each other. The intruder is then stuck there permanently.

Before we decide that interstellar and time travel are categorically impossible consider the miriads of elements, compounds, and substances intersperced in countless forms both animate and inanimate throughout our world. Yet all are composed of the same neutrons, protons and electrons. There is no absolute difference or likeness between anything in the universe. For example, no two protons are exactly alike. Even the same proton is not exactly alike from second to second, if only that it's location changes constantly. Nor is there any absolute difference between solid matter and the fabric of space. In other words there is an inate interchangeability between everything in the universe, including one's position in space and time.

The universe acts as a single entity that is subject to quantum entanglements, nonlocal interactions, and exchanges (time travel). The initial impulse upon breaking time symmetry is described as a transigent action. It has the traveller splinter off of and vector more or less perpendicular to the time continuum. From here the mission will continue on course and enter a parallel universe, or double back to the home universe in order to explore a planet, solar system, or a point in the past or future. Where ageing is the verticle that follows our time arrow, transigent action has the traveller resting between the ticks of the clock while moving horizontally across the greater time plane of the fifth dimension. These are the basic interstellar and time travel logistics.

SIX
MASS ANTIMASS

Consider the galactic juxtapositioning that finds us in what is termed the "local grouping". This includes the Milky Way, Andromeda, and a number of other gravitationally intertwined galaxies. If this pattern is repeated within the greater macrocosm beyond the boundaries of our universe then we, likely, will find communities of other universes similarly transposed. For argument's sake we'll say that there are a hundred of these in that specific cosmic neighborhood of which our universe is one.

There may be inherent differences at the most fundamental level between universes. Here a comparative analysis might reveal whether the laws of physics are identical, similar, different, or diametrically opposed between any and all of these one hundred universes. Will the matter-energy from another such world be compatible or incompatible with that of ours? Can they coexist or anhialate each other the way that matter antimatter does. It is a fair assumption concerning the differences and similarities between foreign worlds that all of the above apply. And that only the statistical odds and specific ratios for each standard model remain in doubt. Because the laws of physics are uniform throughout our universe there may be quantum entanglement as well as gravitational interaction between neighboring galaxies. And matter taken from anywhere else in the cosmos is identicle to that found here on earth. Not so with universes operating under a different set of laws.

Suppose that a small object leaves it's parent universe and takes a course headed for our's. It arrives here, enters the window of your office, and lands on the desk. For obvious reasons anyone in the room at that time will had to have said their prayors in advance of the intruder's arrival. Contact between the object and physical matter will cause the destruction of both. The only question is how deep the dissolution goes, whether it is limited to the molecular structure of matter, or the reaction reaches the atomic or nuclear levels. If the object had been composed of antimatter we'd witness an explosion that would dwarf the detonation of any manmade nuclear device. But this is not the final step in the process. There is an even worse possible outcome; far worse! The end game comes when the energy of the opposing parties also anhialates and disseminates back into the fabric of space. Complete and total anhialation, in other words.

Because matter and antimatter are not diametrically opposed they cannot anhialate completely. To facilitate that type of cataclysm matter from another universe, where the laws of physics oppose our's, must be introduced into this world. For lack of better terminology we'll refer to this correlation as mass antimass. Suppose that a planet sized body breaks away from it's antimass universe and enters ours where, eventually, it collides with a celestial body of similar size in the Milky Way. What we initially see would mimic matter antimatter as it anhialated in a tremendous explosion and release of energy. But then the brilliant glow abruptly turns black as energy anhialates energy and the fabric of space itself explodes. This is where energy is converted back into the fabric of space in accordance to the equation $S=EC^2$ as detailed in chapter one.

Creation is uncreated and absolute blackness ensues. From the earth perspective on a cloudless night we witness a sphere of darkness that snuffs out stars and destroys the planets of our galaxy as it expands faster than the speed of light. The more mass it consumes the more the expansion rate increases, becoming ever more fearsome in an all consuming chain reaction. All discernable form is erased. The Sun and

Earth are not spared. They cease to exist along with the Milky Way. Although the destruction of matter is limited to the Milky Way, the shock wave created by the sudden expansion of space determines that within the local group of neighboring galaxies all gravitationally bound systems will be torn apart or disrupted. Stars, planets, and other celestial bodies are flung out of orbit with many being broken apart. The wave is felt throughout the universe to varying degrees.

Consider the effects on the fabric of space near a blak hole in terms of vector angle and verticle geometry, where objects of matter are torn apart and stretched into thin strands of superheated plasma. Now take into account how much more extreme those conditions become during the mass antimass reaction where the fabric of space explodes.

If there's one saving grace to all of this it's that that reaction unfolds at velocities greater than the speed of light, meaning that it might undergo time reversal which takes it backwards on the time continuum. Therefore, if an antimass body is headed our way (at this time) it may just do an about face into the past upon contact with it's physical counterpart, allowing us to dodge the bullet (in the present). But if it's on a collision course with our future, then we're in big trouble.

Surprisingly, there is one result where if matter from this world and another universe combine they will form a substance that is entirely benign; an end result where no further molecular reaction or interaction is possible. An object of such composition may not be subject to gravitational attraction or even kinetic vectoring and must, therefore, remain in a position of absolute rest. You might refer to it as dysfunctional cosmic sludge frozen somewhere out there in space, just waiting for a conventional celestial body to slam into it.

It is highly unlikely that stars, planets, or celestial objects can break away from their parent universe, and travel all this way in and of themselves. But just as Andromeda is on a collision course with the Milky Way, universe X may also be headed toward us. The question then becomes, is it's matter compatible with with our's? Is it an antimatter or worse yet an antimass universe on course to anhialate with our's?

SEVEN
CORRRECTIONS

Whenever probes are sent out in space to explore one of our neighboring planets, what we discover there is always surprising and generally quite different from what scientists had predicted, in terms of the cold hard facts. The theoretical fields evidence a far greater gap between fact and what is predicted. Theoretical physics is fraught with blind alleys, dead ends, and subject to a myriad of false assumptions. When the subject becomes exceedingly complex such as it did in the book "Beyond the Universe" then errors along the way in persuit of a cohesive theory become even more egregious and must be adressed at some point in the future. Thus, the chapter "Corrections" makes it's appearance.

SUBPARTICLE ASSIMILATION

This was the attempt made, in "counterpart theory", of tying the creation of physical matter-energy to a series of higher dimensional states. The hypothesis proves incorrect, however. Physical matter-energy was not created out of, nor is it subject to dissolution back into the material of any higher dimensional world. Here the dimensional disparity between all such worlds is one prohibiting factor. Physical matter-energy is only compatible with physical matter-energy and solely interchangeable with physical space/time. Thus, the hypothesis detailing "direct translation" is the correct one.

COMPENSATION FACTORS

Light speed is defined as the cooperation between each photon and the fabric of space. If it's velocity is to remain constant then the consistency of the fabric of space must also remain constant and not subject to expansion or contraction. It amounts to an either or proposition where both light speed and space are constant or neither will be. Initially the big bang and expansion of space were accepted as proven fact in counterpart theory. This expansion was explained on the basis of a positive and negative compensation factor inherent to space, under the assumption that that action was entirely self contained within the boundaries of one universe (our's). But that also proved to be untrue. A microbe might suppose that it is riding on the back of an elephant when it is merely sitting on the back of a beetle that is riding on the back of that elephant. Similarly, the dynamic that underlies the expanding universe spans beyond the macrocosm and is, therefore, not explainable from a perspective taken (within) the universe. It can only be understood in terms of that which is being played out in the greater macrocosm of the fifth dimension.

THE BIG BANG

If, in fact, the fabric of space is not undergoing expansion in and of itself, then the entire "big bang" scenerio becomes implausible and we must redefine our cosmological model.

TIME RADIUS POSITIONING

Time dilation as it occurs under acceleration or within a gravitational field has more to do with being in a state of suspended animation than it does with the time dimension itself. Therefore, "time radius positioning" as described in "Beyond the Universe" is not required in explaining the time dilation effect and, as such, becomes a mute point.

EIGHT
FACT VERSUS BELIEF

It generally holds true that each one of us as an individual has a world view. This is a system of knowledge, opinions, and beliefs that allows us the framework for understanding the true nature of the world we live in. It differs for most everyone according to the type of person they are. Whether emotionally or intellectually driven, religeous or atheistic, liberal or conservative, and the type of education and upbringing they have had. One thing holds true for both believer and sceptic alike. No matter if their world contains God, no God, UFOs or weather balloons, big foot or a man in a monkey suit, evolution or creationism; their world view for the most part is not based on the facts or any real proof but, rather, on what they want to believe. That which secures them firmly in their comfort zone, in accordance to any predisposition. For example, emotionally motivated people seek solace in a heavenly father, whereas, intellectuals tend to deny the existence of anything or anyone having a higher degree of intelligence than they do. God, spiritual beings, extraterrestrials, and soforth.

You'll recall the story about captain Smith of the Titanic who obligingly went down with the ship after it had struck an iceberg. Similarly, many of us will continue to cling to our world view long after it has been rendered unseaworthy. Such is the case with climate change denial which is in no way supported by any facts scientific or other-wise. Instead, it is motivated solely by corporate profit margins and political affiliation. If allowed to do so these deniers will continue their untenable environmental policies until the bitter end, no doubt, with the inevitable collapse of this planet's ecosystem and the extinction of mankind. You can think of it as the captain Smith syndrome gone mad as humanity goes down with the ship on a worldwide scale. This type of false world view can persist only where the facts are adjusted to fit the individual's personal bias; what one wants to believe as opposed to the way it really is. And it is not limitted solely to those who are naive enough to fall into the climate denial category. Personal bias can also taint science at the highest level. There are scientists who have spent entire careers persuing a certain set of beliefs such as 'string theory'. Can you blame them for fighting tooth and nail to salvage an untenable theory when the alternative is to admit that your life's work had been a failure. To either fail or go down with the ship, those are the options.

Is it coincidental that children requiring a Santa Clause to bring presents at Christmas mirrors physicists needing a Higg's boson to explain mass in the universe? Here, what one wants to believe may trump that which is actually true. Those seeking proof of the mysterious "God particle" will misinterpret test data and skewer the facts to insure the desired results, that the correct particle has been found and identified as predicted. In response to that effort one can only paraphrase a quote from President Obama "you can put all the lipstick you want on the pig, just don't call it Miss America". Mass is a characteristic that is intrinsic to all discernable form in the universe. No mysterious agent is required in manifesting it any more than we need a Santa Clause to account for our Christmas gifts.

Even our most celebrated genious, Albert Einstein, had a personal bias against the randomness at the quantum level that had been implicated by Heisenberg. "God does not play dice with the universe" was his response to proponents of "Quantum mechanics".

Consider that certain world views seem to have been custom fit for their proponents to the degree of an Armani suit. For example, the idea has been proposed that our universe is the manifestation of a giant computer program. For obvious reasons this type of cosmic model can only be the brainchild of your consumate technobuff, albeit a very creative one. And it would surely constitute 'paradise on earth' for anyone so predisposed. Computer generated universe, accurate description, or bias view as the result of wishful thinking?

It seems that we have two opposing phylosophies regarding world views. The first states emphatically "this is all there is. There is nothing beyond the physical world". To which the second view counters that "this can't possibly be all there is. There must be something greater than physical reality". Thus, one must choose sides carefully (via the toss of a coin) and defend it vociferously. "Not a shred of evidence" is the preferred disclaimer used by sceptics in dismissing various phenomenon that conflict with their preferred world view. Where UFOs are concerned they can barely navigate past the mountain of evidence via that tactic. However, when adressing spiritual and other worldly manifestations the sceptics are absolutely correct in their assertions. No evidence has been found nor will it ever be. Searching for proof of spiritual phenomenon in the gross matter of the physical world is as futile an attempt as looking for camels in the ocean. You could sail the seven seas for a lifetime and not find a single one, then likely conclude that there's no such thing. Other worldly manifestations will not budge the needle one iota on the physical instruments used to detect them. The effect is similar to neutrinos passing through blocks of lead unhindered. The dimensional disparity between physical and non-physical matter makes any interaction between the two even more unlikely.

No public disinformation campaign has ever been initiated against Santa Clause, neither has anyone tried to debunk the Easter Bunny. The reason why is that, for the most part, only actual phenomenon requires debunking. The fact that such a concerted and vehement effort has been made in discrediting UFO witnesses only adds more credence to the phenomenon. If there'd been nothing to it in the first place, it would have been nonchalantly dismissed at the outset and forgotten about.

As a general rule members of the scientific community in good standing will obligingly adhere to the "reductionist theory", at least on the record. What this states is that all events in the universe can be explained on the basis of molecular activity. In fact, not even all molecular action can be explained on the basis of molecular activity and, upon entering the quantum level it becomes even more chaotic and unexplainable.

Reductionism is an atheistic and simplistic way of viewing the world, and especially ourselves. It denigrates us as living beings to think that our life, our intelligence, our very being is nothing more than a byproduct of molecular activity. When we reach the crossroads where reductionism collides with religeon, then we are faced with an either or proposition that allows for no shades of gray. Either life is initiated at conception through molecular activity; or it is introduced into the body as soul from a higher spiritual state. Those are the only two options available to us, no mix and match or combination of the two is allowed here. Take your pick. Also, if soul originates in a higher dimensional state then it is not bound by the limitations of physical clocks and calendars, nor is it created or uncreated within those durational parameters. In other words soul pre-exists the physical body as well as surviving it's death.

Suppose that there is another earthlike habitable planet somewhere else in our galaxy where life had been created and evolved along parallel lines to our own. Plants, animals, insects, and even people all thrive, with one exception. Flight had never been developed there, no birds, or flying insects. Only the ground dwelling varieties of each species can be found in that ecosystem. We'll date this story at some time in the future well after man had perfected interstellar flight.

Scientists discover this world, we'll call planet X, and subsequently send a large mission there designed to conduct a long term experiment. A scientific research center and industrial complex are constructed. The site is manned with a crew of scientists, engineers, and manufacturers who are native to planet X. They are given unlimitted resources and told to research and develope any and all cutting edge technology to advance their civilization. The question to be answered then is whether they will be able to concieve of and engineer a flying machine of some sort, and the timeline involved. It's likely they will, being the brightest minds to be found there, but it should take quite a bit longer than it had on earth. Where the Wright brothers had the benefit of studying birds in flight, the planet X crew cannot do so. Orville and Wilbur Wright needed only to bridge the gap between biological and mechanical flight. On planet X people lack the 'conceptual basis' for any kind of flight and must, therefore, go from no flight to mechanical flight. Instead of the hundred thousand plus years it took earthlings to perfect flight, it might take planet X over a million years to do so. Now to phase two of the experiment. We'll remove the entire crew of scientists and engineers, replacing them with an equal number of chimpanzees. Now how long will it take to engineer flight? Never. And why not? Too dumb, the chimps don't have the intelligence to conceptualize or create anything, let alone something as sophisticated as an aircraft. For the final part of the experiment we'll dumb down even more. The chimpanzees are removed leaving only the raw materials and machinery to manufacture an aircraft. The entire industrial complex is then placed on an earthquake simulator and set on high for a specified time. After shaking for a million or so years we shut it off, open the hangar and, voila, there sits a fully assembled, functional 747 ready to roll down the runway. Ridiculous scenerio? Certainly. No one would be foolish enough to believe that a jet aircraft could be randomly created with no intelligent input. But this is exactly what conventional science tells us about biological flight and birds; that they were conjured up by our "dumb universe" without the benefit of any intelligent thought, planning, or action. Obviously, the reductionist picture gives us a naive over-simplification of our world.

At the opposite end of the scale we must adress the worldview of traditional religeon. We are told that the supreme being created our universe lock, stock, and barrel in and of his own power; and seemingly out of nothing in hardly any time at all. But this is and inaccurate and over-simplified description of the relationship that God would have with the physical world. If the hypothesis for subparticle assimilation is untrue, then the existence of the universe cannot be backtracked to a series of constructs involving God or the higher dimensional states. Physical reality and it's matter-energy has a validity in and of itself apart from any creationist or big bang scenerio, where something is conjured up out of nothing.

You may go to the beach and build sand castles but you certainly cannot say that you had created sand or the beach. Your input there was limited solely to rearranging the sand that had already been there. This is very similar to intelligent input on a cosmic scale with the key term being input as opposed to creation. Just as the man on the beach did not create the sand the higher intelligence did not create the universe or it's matter-energy. Instead, you might say that God exercises concious control over our world and without that input planet Earth would be as lifeless and barren a place as Mars.

The physical universe, parallel universes, higher and lower dimensional worlds all exist in a simultaneous state independant of each other. There is no creation scenerio or cause and effect sequence linking any of them together in the manner that you would see between the chicken and the egg, where one exists as the result of the other.

God, super computer program, various spiritual beings, higher intelligence from extraterrestrial sources, or parallel worlds. These are some of the options for pinning down the "plus element" at work in our universe, The source and definition of that intelligence will be a matter of opinion and personal preference for each individual. Mankind stands at the apex of all things manifest in the universe and may thus have kinship to it.

Consider the cariovascular system, central nervous system, digestive system, the skeleton, the eyes and ears, and the brain. The notion that each of these complex biosystems, critical to human life can be created and perfected, working in unison with each other as the result of countless evolutionary rolls of the dice taken by our 'dumb universe' with no intelligent input is the prime example where reductionist dogma trumps fact and common sense. Our entropy prone universe would be less capable of creating the functional complexity of a concious human being than that human is capable of stirring the sugar cube back out of his morning cup of coffee. Without the intelligent input, left totally to chance, and random action, life would de-evolve backwards into extinction.

Now to properly conclude this study let us recite "the sceptic's prayer". Thank you Lord for withholding the intelligent input when you created us and substituting it instead with swamp gas and hypnogogic delusions. We thank thee for giving us weather balloons and the light of Venus, without which we wouldn't have a leg to stand on in our endless battle against the hated UFOlogist. And when they no longer believed the weather balloon story you blessed us with the new and improved project mogul balloons. Thank you Lord for the unreliability of eyewitness testimony and the campaign of disinformation issued by the United States Military. We thank thee for allowing us to be so far in denial that we can even deny being in denial. Let us remember Roswell and the Patterson Gimlin film. Blessed is the man in the monkey suit and the crash test dummy that fell out of his weather balloon while chasing the light of Venus. We are so blessed in the belief that the physical world is all there is, that when we die we won't have to ascend up to heaven to live forever in paradise with thee. Instead, we can be buried underground where we'll decompose and get eaten up by worms. Lead us not into close encounters of the third kind, and deliver us from Roswell and the rantings of professor Stanton Friedman. In Joe Nichols name, amen.

Roswell

This just in: The real story behind the 1947 incident is finally revealed.

Hoaxes, temperature inversions, and the light of Venus notwithstanding, the crash test dummy fell out of his weather balloon after inhaling too much swamp gas and suffering from hypnogogic delusions as the result of the military's disinformation campaign but, on the other hand, we realize just how unreliable eyewitness testimony really is.

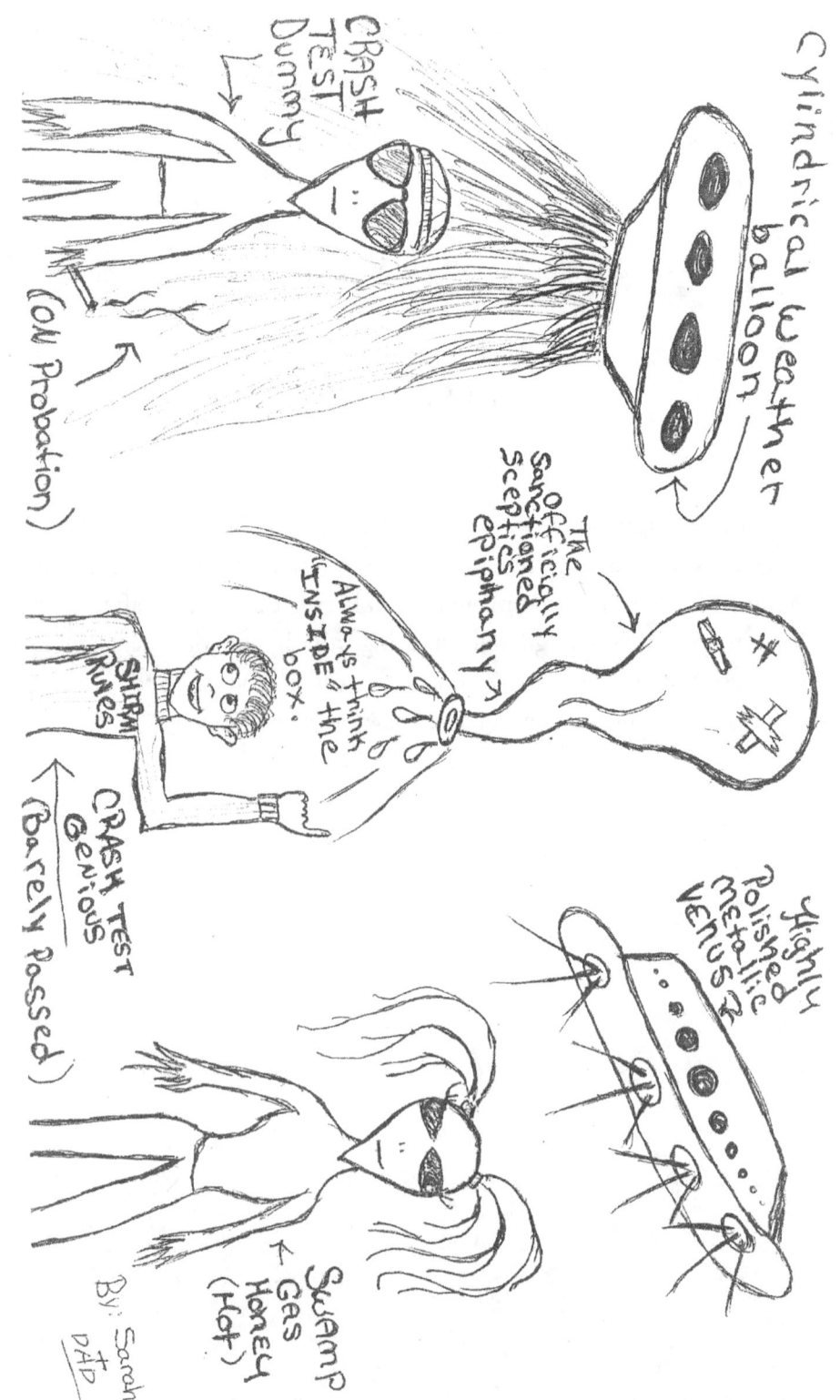

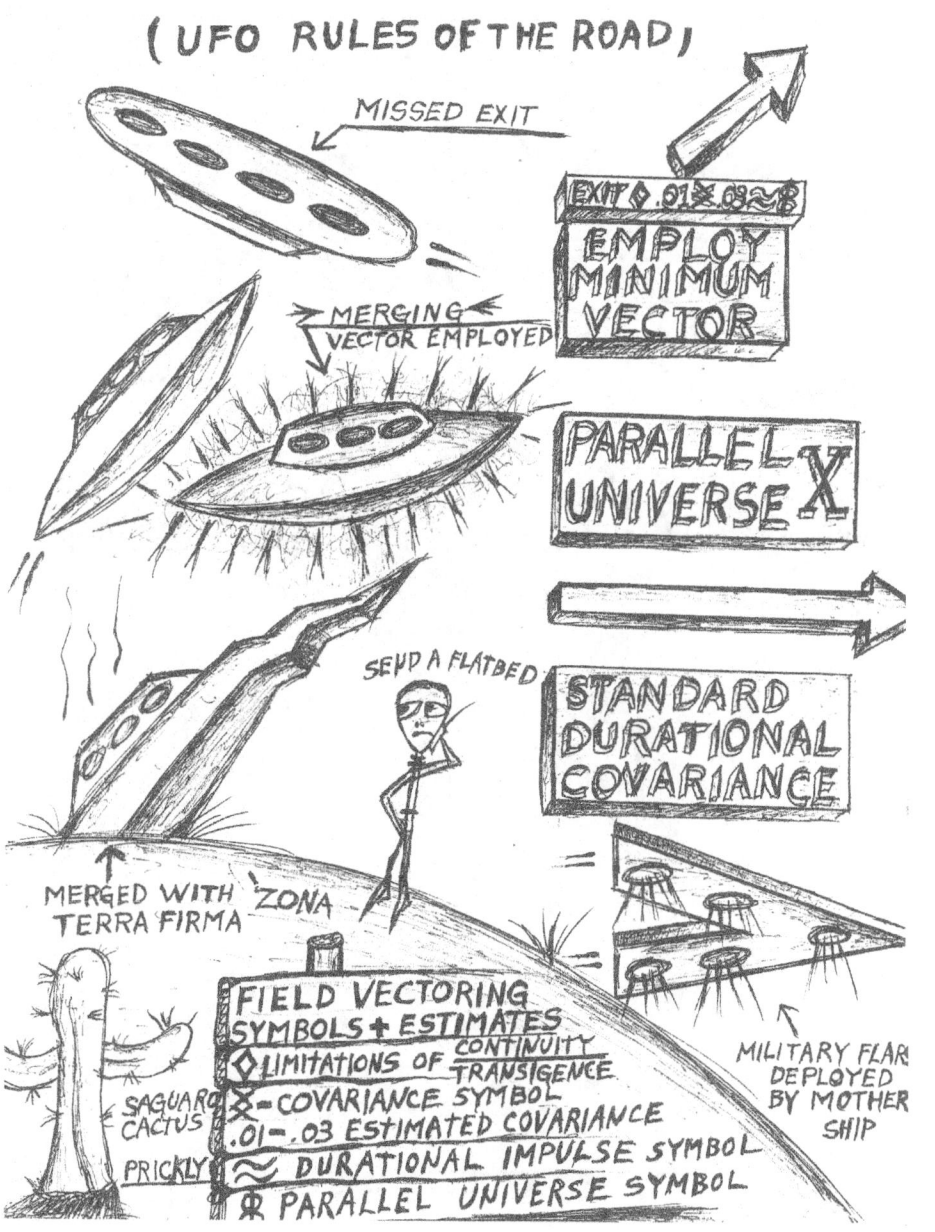

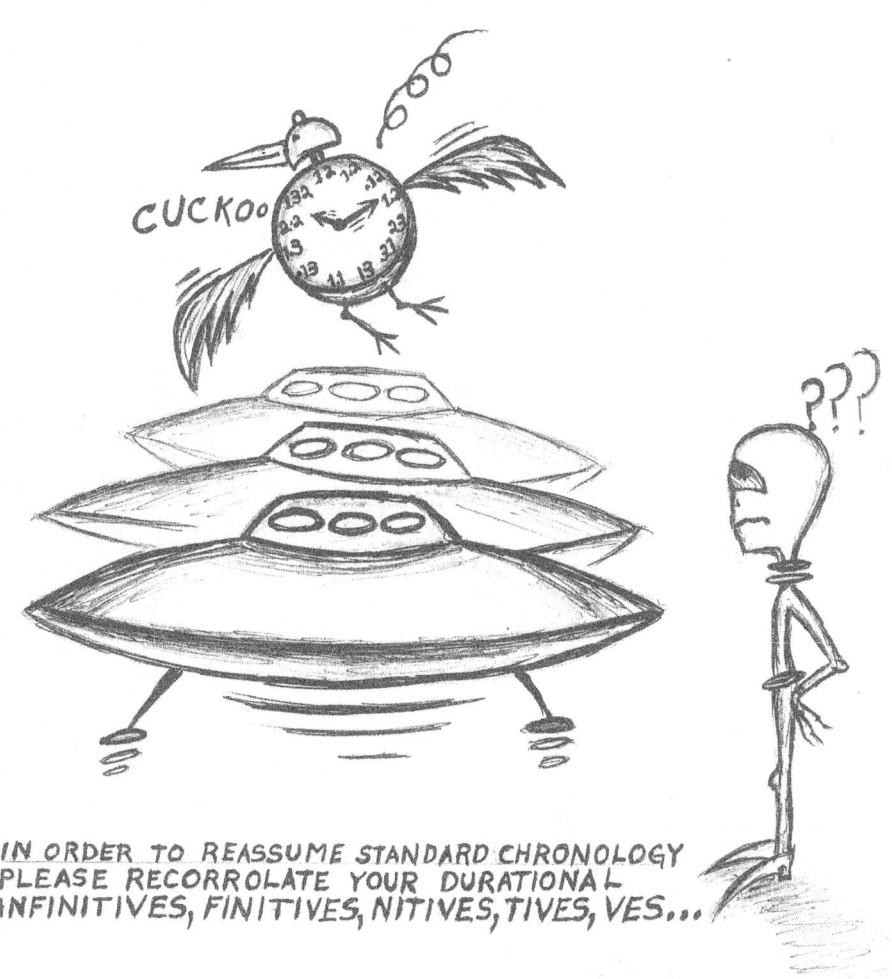

www.ingramcontent.com/pod-product-compliance
Lightning Source LLC
Chambersburg PA
CBHW061518180526
45171CB00001B/233